CHATGPT FOR CASH FLOW

10 EASY WAYS TO UNLOCK THE POWER OF AI TO BUILD A SIDE HUSTLE EMPIRE & MAKE MONEY ONLINE FAST

MARK SILVER

ChatGPT for Cash Flow: 10 Easy Ways to Unlock the Power of AI to Build a Side Hustle Empire & Make Money Online Fast

EMPIRE Publishing

Houston, TX

www.empireghostwriter.com/book-process

contact@empireghostwriter.com

@empirepublishing

This book is not meant to serve as financial, investment, or business advice, but merely for educational purposes.

ISBN [ebook] - 978-1-956283-60-0
ISBN [paperback] - 978-1-956283-61-7
ISBN [hardcover] - 978-1-956283-66-2

CONTENTS

INTRODUCTION

In today's fast-paced and ever-evolving world, financial stability and independence are becoming increasingly important. The rise of the gig economy has led to an explosion of side hustles, or part-time ventures that people undertake in addition to their regular jobs to earn extra income.

These side hustles not only help individuals supplement their income but also provide opportunities to explore passions, hone skills, and potentially transition into a full-time career.

One revolutionary tool that has emerged this year is ChatGPT, an advanced language model developed by OpenAI. ChatGPT has the ability to generate human-like text based on given inputs, opening up a world of possibilities for entrepreneurs, freelancers, and side hustlers. By leveraging ChatGPT, you can boost your productivity, streamline your workflow, and expand the range of services you offer.

In this book, ***ChatGPT for Cash Flow: 10 Easy Ways to Unlock The Power of AI to Build a Side Hustle Empire & Make Money Online Fast***, we will explore the benefits of using ChatGPT for side hustles, discuss the criteria for selecting the top 10 side hustles, and delve into each hustle in detail. We will provide practical tips and guidance on how to get started with each side hustle and how to maximize your earnings potential using ChatGPT.

Whether you are a freelancer, a small business owner, or simply someone looking to make some extra money on the side, this book will equip you with the knowledge and resources to leverage the power of ChatGPT and succeed in your online ventures. So, let's embark on this exciting journey to discover the top 10 side hustles that can help you achieve financial independence and success.

WHAT ARE SIDE HUSTLES AND WHY ARE THEY IMPORTANT?

Side hustles are supplementary income-generating activities that people engage in apart from their primary occupation. These activities can include a wide range of pursuits, such as freelance work, selling products or services, or participating in the gig economy.

Side hustles hold significant importance for several reasons.

1. **Financial security:** Engaging in side hustles provides an additional income stream, helping individuals achieve better financial stability. This extra money can be used to pay off debts, build emergency funds, or invest in future goals.
2. **Skill development:** Pursuing side hustles allows people to develop and refine new skills, benefiting them personally and professionally. Acquiring these skills can lead to career growth, improved job opportunities, and increased job security.
3. **Entrepreneurship opportunities:** Side hustles offer a low-risk way to experience entrepreneurship. Many successful businesses have started as side projects, and engaging in a

side hustle can provide valuable insights, connections, and experiences that could eventually evolve into a full-time venture.

4. **Improved work-life balance:** Side hustles often grant more flexibility and control than traditional jobs, enabling individuals to strike a better work-life balance. They can serve as an outlet for pursuing passions, interests, or hobbies that might not be possible with a full-time job.
5. **Networking and personal development:** Diving into side hustles can lead to new connections, collaborations, and opportunities for personal growth. Networking with like-minded individuals can unlock new possibilities and expand one's professional circle.

In summary, side hustles are essential because they offer financial advantages, skill development, entrepreneurial opportunities, a better work-life balance, and chances for networking and personal growth. By engaging in side hustles, individuals can create a more rewarding and secure professional journey.

CHATGPT AND SIDE HUSTLES: A WINNING COMBINATION FOR ONLINE SUCCESS

ChatGPT is an innovative language model developed by OpenAI that harnesses the power of artificial intelligence to generate human-like text based on given inputs. It can understand context, produce coherent responses, and create high-quality content, making it a valuable tool for a wide range of applications.

In the context of side hustles, ChatGPT can be a game-changer by enhancing productivity, streamlining workflows, and expanding the scope of services you offer. Here are some ways it can be utilized for various side hustles.

1. **Content creation:** ChatGPT can generate well-written articles, blog posts, and social media content, enabling you to offer content creation services to clients or improve your online presence.
2. **Copywriting:** ChatGPT can assist in crafting persuasive copy for advertising, marketing, and sales materials, making it easier for you to offer copywriting services to businesses and entrepreneurs.
3. **Research and summarization:** ChatGPT can help gather information and summarize it in a

clear, concise manner, which can be beneficial for creating reports, whitepapers, or even researching topics for your projects.

4. **Virtual assistance:** ChatGPT can handle various administrative tasks, such as email drafting, appointment scheduling, or data entry, allowing you to offer virtual assistance services to clients in need.
5. **Creative writing:** Whether it's writing children's books, short stories, or poetry, ChatGPT can provide creative inspiration and help with the writing process, enabling you to explore new avenues in creative writing.
6. **Language translation:** ChatGPT can assist with translating text between multiple languages, opening up opportunities to offer translation services to a global clientele.

By leveraging ChatGPT for your side hustles, you can not only save time and effort but also improve the quality and range of services you provide. These results can lead to increased earnings, a more extensive client base, and a competitive edge in the ever-growing gig economy.

1

WRITING FOR BLOGS AND WEBSITES WITH CHATGPT

In this chapter, we will discuss how writing for blogs and websites can be a profitable side hustle and how ChatGPT can assist with research and writing. We will also cover how to find clients and market your services, as well as give tips for maximizing your earnings potential.

WRITING FOR BLOGS AND WEBSITES AS A SIDE HUSTLE

Writing for blogs and websites is an excellent side hustle option for several reasons.

First, there is a high demand for quality content. Businesses and individuals alike require well-written articles to engage their audience, improve their search engine rankings, and establish their expertise.

Second, freelance writing offers flexibility, allowing you to work on your schedule and choose projects that align with your interests and expertise.

Finally, it requires minimal upfront investment, with only a computer and an internet connection needed to get started.

The Benefits and Drawbacks of Freelance Writing

Benefits

- **Flexibility:** Work on your schedule and set your hours.
- **Location independence:** Write from anywhere with an internet connection.
- **Diverse projects:** Choose from various topics and industries to keep things interesting.
- **Skill development:** Improve your writing skills and learn about new subjects.

Drawbacks

- **Inconsistent income:** Earnings may vary month-to-month.

- **Finding clients:** Building a client base can be challenging.
- **Self-discipline:** Staying motivated and managing your time effectively is essential.

How Much Can You Earn as a Freelance Writer?

Freelance writing income can vary significantly depending on factors such as experience, niche, and client base. On average, beginners may earn between $25-$50 per article, while experienced writers can charge $100-$500 or more per piece. Some writers may also charge per hour or per word.

HOW CHATGPT CAN ASSIST WITH RESEARCH AND WRITING

ChatGPT can assist freelance writers with various tasks, such as researching topics, generating article outlines, suggesting content ideas, and even writing drafts.

The Features and Benefits of ChatGPT for Freelance Writers

Features

- **Research assistance:** Quickly gather information on a wide range of topics.

- **Content generation:** Generate outlines, ideas, or drafts to kickstart your writing.
- **Language optimization:** Improve your writing by suggesting better word choices or phrasing.

Benefits

- **Time savings:** Streamline your research and writing processes.
- **Enhanced creativity:** Generate fresh ideas and perspectives for your content.
- **Increased productivity:** Produce more high-quality content in less time.

How to Use ChatGPT to Streamline Your Writing Process

- **Start with a prompt:** Give ChatGPT a specific question or topic to research.
- **Review and refine the output:** Use the generated content as a starting point and tailor it to your voice and style.
- **Optimize your writing:** Leverage ChatGPT's suggestions to improve your text's clarity and readability.

CHATGPT IN ACTION

Use the following prompt to gather blog topic ideas from ChatGPT:

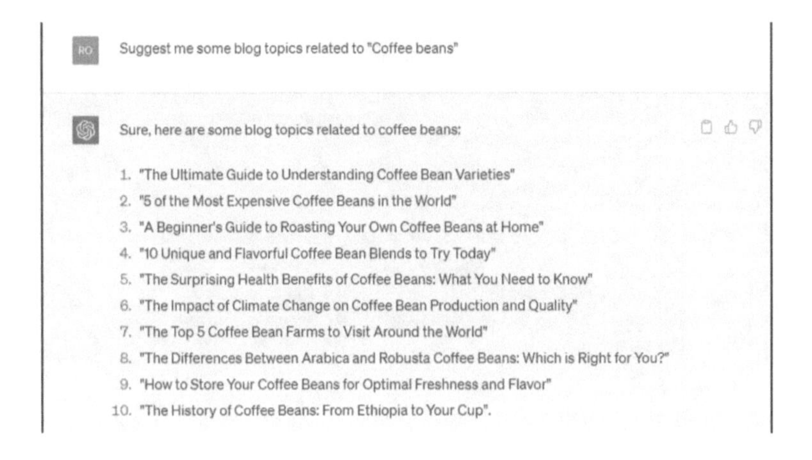
RO Suggest me some blog topics related to "Coffee beans"

Sure, here are some blog topics related to coffee beans:

1. "The Ultimate Guide to Understanding Coffee Bean Varieties"
2. "5 of the Most Expensive Coffee Beans in the World"
3. "A Beginner's Guide to Roasting Your Own Coffee Beans at Home"
4. "10 Unique and Flavorful Coffee Bean Blends to Try Today"
5. "The Surprising Health Benefits of Coffee Beans: What You Need to Know"
6. "The Impact of Climate Change on Coffee Bean Production and Quality"
7. "The Top 5 Coffee Bean Farms to Visit Around the World"
8. "The Differences Between Arabica and Robusta Coffee Beans: Which is Right for You?"
9. "How to Store Your Coffee Beans for Optimal Freshness and Flavor"
10. "The History of Coffee Beans: From Ethiopia to Your Cup".

Use the following prompt to generate the article with ChatGPT:

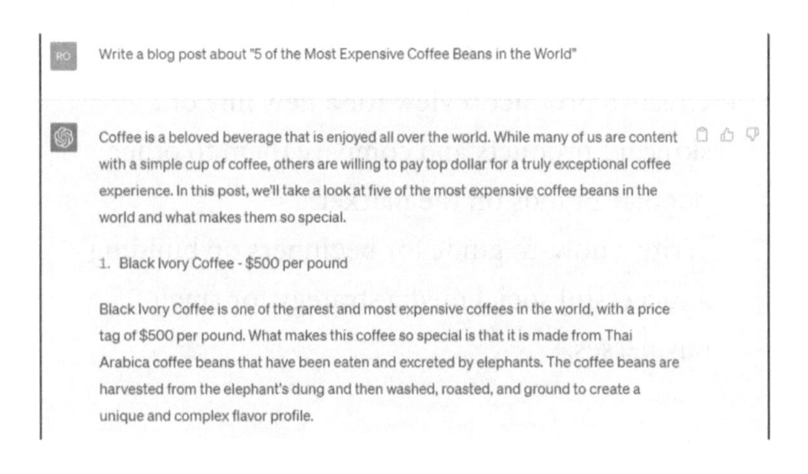
RO Write a blog post about "5 of the Most Expensive Coffee Beans in the World"

Coffee is a beloved beverage that is enjoyed all over the world. While many of us are content with a simple cup of coffee, others are willing to pay top dollar for a truly exceptional coffee experience. In this post, we'll take a look at five of the most expensive coffee beans in the world and what makes them so special.

1. Black Ivory Coffee - $500 per pound

Black Ivory Coffee is one of the rarest and most expensive coffees in the world, with a price tag of $500 per pound. What makes this coffee so special is that it is made from Thai Arabica coffee beans that have been eaten and excreted by elephants. The coffee beans are harvested from the elephant's dung and then washed, roasted, and ground to create a unique and complex flavor profile.

Use the following prompt to proofread the article with ChatGPT:

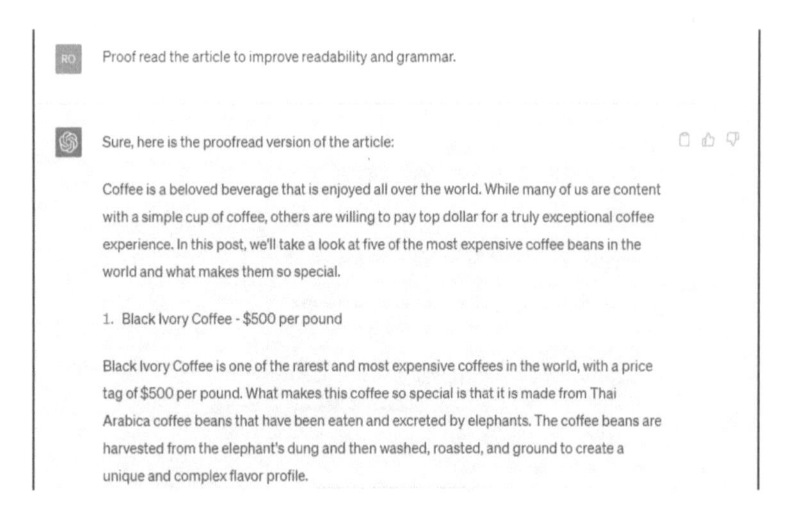

PROMPTS FOR BLOG AND CONTENT WRITING SIDE HUSTLE

- Write a blog post about the benefits of mindfulness meditation for mental health.
- Create a product review for a new line of skincare products and compare them to other popular brands on the market.
- Write a how-to guide for beginners on building a successful social media strategy for small businesses.

- Create a listicle of the top 10 travel destinations to visit in 2023 and why they should be on your bucket list.
- Write a personal essay on the challenges and rewards of working from home in today's digital age.
- Create a buying guide for eco-friendly products and the benefits of choosing sustainable options.
- Write a feature article on the rise of veganism and its impact on the food industry.
- Write an About Us page that tells the story of your company and why you started it.
- Create a landing page for a new product or service that showcases its features and benefits.
- Write an FAQ page that answers common questions about your company, products, and services.
- Create a blog page that highlights your company's expertise and industry knowledge.
- Write a product page that provides detailed information on each of your products, including pricing and specifications.
- Create a testimonial page that showcases customer reviews and feedback.
- Write a page that outlines your company's values and mission statement.

- Create a careers page that lists job openings and describes the company culture.
- Write a contact page that includes a contact form, phone number, and email address for customer inquiries.
- Create a resources page that offers helpful articles, guides, and resources for your audience.

FINDING CLIENTS AND MARKETING YOUR SERVICES

How to Find Clients as a Freelance Writer

- **Job boards:** Search for freelance writing opportunities on platforms like Upwork, Freelancer, and ProBlogger.
- **Cold pitching:** Reach out to businesses or publications in your niche to offer your writing services.
- **Networking:** Attend industry events, join online communities, and connect with fellow writers to discover potential clients.

The Importance of Building a Portfolio and Networking

A portfolio showcases your writing skills and expertise, helping potential clients assess your suitability for their projects. To create an effective portfolio, include a variety of your best work, ideally within your chosen niche.

You can also create a personal website or use portfolio platforms, like Medium or Journo Portfolio, to showcase your writing samples.

Networking is vital for finding clients and building long-term relationships. Connect with other writers, editors, and content managers through online forums, social media platforms, and professional networking sites like LinkedIn.

By engaging with others in the industry, you'll not only discover potential clients but also learn valuable insights and stay up-to-date with industry trends.

TIPS FOR MARKETING YOUR SERVICES EFFECTIVELY

- **Develop a personal brand:** Differentiate yourself from other writers by highlighting your unique voice and expertise.
- **Utilize social media:** Share your work and engage with your target audience on platforms like Twitter, LinkedIn, and Facebook.

- **Guest post:** Offer to write guest articles for reputable blogs and websites within your niche to gain exposure and credibility.
- **Offer value:** Share valuable content, insights, and tips within your industry to attract potential clients.

MAXIMIZING YOUR EARNINGS POTENTIAL

Strategies for Increasing Your Rates and Earning More

- **Specialize in a niche:** Writers with niche expertise can often charge higher rates due to their specialized knowledge.
- **Improve your skills:** Invest in professional development through courses, workshops, and books to become a better writer and increase your value.
- **Upsell related services:** Offer additional services like content editing, SEO optimization, or social media management to increase your earnings.
- **Obtain retainer agreements:** Establish ongoing relationships with clients who require regular content, providing a stable income source.

How to Negotiate with Clients

- **Know your worth:** Research industry rates and understand the value of your skills and expertise.
- **Be confident:** Clearly communicate your value proposition and justify your rates with examples of your work and results.
- **Offer tiered pricing:** Provide clients with multiple pricing options, allowing them to choose the level of service that fits their needs.
- **Be flexible:** Be open to negotiation but know your limits, and be prepared to walk away from a project if the terms don't meet your requirements.

The Importance of Setting Goals and Tracking Your Progress

Setting clear goals helps you stay focused and motivated in your freelance writing business. Establish both short-term and long-term objectives, such as monthly income targets or the number of clients you want to acquire. Regularly track your progress and adjust your strategies as needed to stay on track toward your goals.

Writing for blogs and websites can be a lucrative side hustle with the right approach. By leveraging the power of ChatGPT and following these tips, you can build a successful freelance writing business and achieve your financial goals. Stay persistent, keep refining your skills, and market your services effectively to unlock your full potential as a freelance writer.

2

VIDEO SCRIPT AND DESCRIPTION WRITING WITH CHATGPT

In the age of digital media, video content has become increasingly important for businesses, influencers, and individuals alike. As a result, the demand for skilled video script and description writers has grown.

This chapter will provide an overview of video script and description writing as a side hustle, explain how ChatGPT can assist you in this field, guide you on finding clients and marketing your services, and offer tips to maximize your earnings potential.

VIDEO SCRIPT AND DESCRIPTION WRITING AS A SIDE HUSTLE

A. Video Script Writing

Video script writing involves crafting engaging and informative scripts for various types of video content, such as commercials, tutorials, product reviews, documentaries, and more. Effective scripts not only convey the intended message but also create a connection with the target audience.

B. Video Description Writing

Video description writing focuses on creating concise and accurate summaries of video content. These descriptions help viewers understand the content of a video before watching it and are essential for search engine optimization (SEO).

Why Is Video Script and Description Writing a Great Side Hustle Option?

Video content has exploded in popularity, with businesses and individuals leveraging it for marketing, education, and entertainment. Writing video scripts and descriptions is an excellent side hustle option, as it requires creativity, attention to detail, and a knack for storytelling. Plus, it's an exciting and engaging form of writing that allows you to flex your creative muscles.

The Benefits and Drawbacks of Video Script and Description Writing

Benefits

- **Creative expression:** Write engaging and compelling stories.
- **Flexibility:** Work on your schedule and choose projects that align with your interests and expertise.
- **Continuous demand:** The demand for quality video content is growing.
- **Potential for long-term projects:** Clients may require ongoing video content.

Drawbacks

- **Requires some technical knowledge:** You may need to understand video production and editing to write compelling scripts and descriptions.
- **Can be challenging to break into:** Establishing yourself in the industry may take time and effort.

How Much Can You Earn as a Video Script and Description Writer?

As with any freelance writing gig, rates can vary based on experience, niche, and project scope. On average, beginner writers may earn between $50-$100 per minute of video content, while experienced writers can charge $200 or more per minute.

HOW CHATGPT CAN ASSIST WITH RESEARCH AND WRITING

How ChatGPT Can Help with Research and Writing

ChatGPT can assist video script and description writers in various tasks, such as researching topics, generating video outlines, suggesting content ideas, and even writing drafts.

The Features and Benefits of ChatGPT for Video Script and Description Writers

Features

- **Research assistance:** Quickly gather information on a wide range of topics.
- **Content generation:** Generate outlines, ideas, or drafts to kickstart your video writing.

- **Language optimization:** Improve your writing by suggesting better word choices or phrasing.

Benefits

- **Time savings:** Streamline your research and writing processes.
- **Enhanced creativity:** Generate fresh ideas and perspectives for your video content.
- **Increased productivity:** Produce more high-quality video content in less time.

How to Use ChatGPT to Streamline Your Writing Process

- **Start with a prompt:** Give ChatGPT a specific question or topic to research.
- **Review and refine the output:** Use the generated content as a starting point and tailor it to your voice and style.
- **Optimize your writing:** Leverage ChatGPT's suggestions to improve your script's clarity and pacing.

CHATGPT IN ACTION

Use the following prompt to generate Nas Daily Style YouTube video:

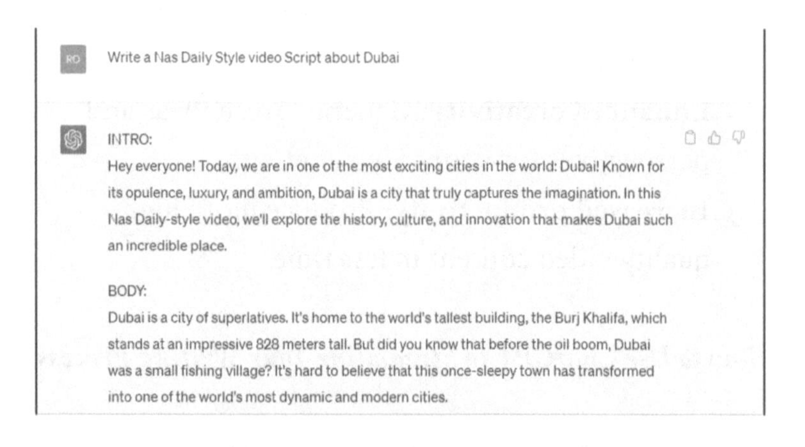
RO Write a Nas Daily Style video Script about Dubai

INTRO:
Hey everyone! Today, we are in one of the most exciting cities in the world: Dubai! Known for its opulence, luxury, and ambition, Dubai is a city that truly captures the imagination. In this Nas Daily-style video, we'll explore the history, culture, and innovation that makes Dubai such an incredible place.

BODY:
Dubai is a city of superlatives. It's home to the world's tallest building, the Burj Khalifa, which stands at an impressive 828 meters tall. But did you know that before the oil boom, Dubai was a small fishing village? It's hard to believe that this once-sleepy town has transformed into one of the world's most dynamic and modern cities.

Use the following prompt to generate Mr. Beast Style YouTube video:

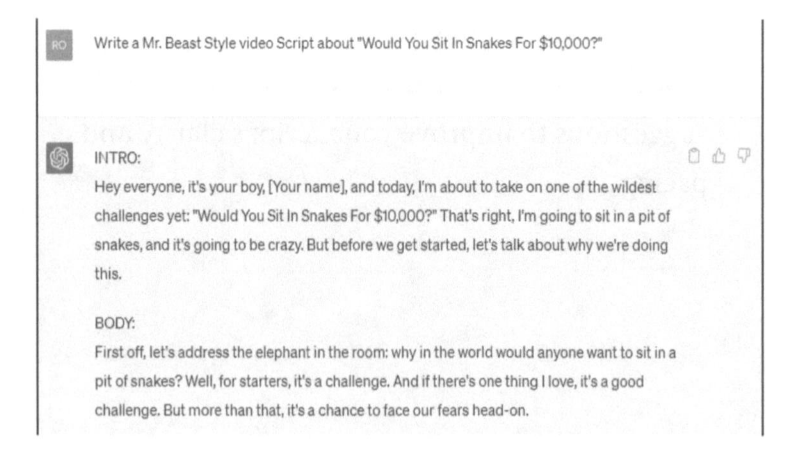
RO Write a Mr. Beast Style video Script about "Would You Sit In Snakes For $10,000?"

INTRO:
Hey everyone, it's your boy, [Your name], and today, I'm about to take on one of the wildest challenges yet: "Would You Sit In Snakes For $10,000?" That's right, I'm going to sit in a pit of snakes, and it's going to be crazy. But before we get started, let's talk about why we're doing this.

BODY:
First off, let's address the elephant in the room: why in the world would anyone want to sit in a pit of snakes? Well, for starters, it's a challenge. And if there's one thing I love, it's a good challenge. But more than that, it's a chance to face our fears head-on.

PROMPTS FOR VIDEO SCRIPT AND DESCRIPTION WRITING

- Create a video script and description for a product launch video that showcases the features and benefits of a new fitness tracker.
- Write a video script and description for a brand story video that highlights the history and mission of a sustainable fashion brand.
- Create a video script and description for a tutorial video that teaches viewers how to make a DIY home decor project.
- Write a video script and description for an explainer video that breaks down the science behind climate change and its effects on the environment.
- Create a video script and description for a promotional video that promotes a new travel destination and highlights its unique attractions.
- Write a video script and description for a how-to video that teaches viewers how to cook a new recipe from scratch.
- Create a video script and description for an educational video that explores the history and cultural significance of a specific type of music.

- Write a video script and description for a testimonial video that features real customers sharing their experiences and feedback about a product or service.
- Create a video script and description for an interview video that features an expert in a specific field sharing their knowledge and insights.
- Write a video script and description for a vlog-style video that documents a day in the life of a professional in a specific industry.

FINDING CLIENTS AND MARKETING YOUR SERVICES

How to Find Clients as a Video Script and Description Writer

- **Freelancing websites:** Create profiles on popular freelance websites, such as Upwork, Fiverr, and Freelancer, to showcase your skills and portfolio.
- **Cold pitching:** Reach out to businesses or video production companies to offer your writing services.

- **Networking:** Attend industry events, join online communities, and connect with fellow video creators to discover potential clients.
- **Social media:** Join niche-specific groups and forums on platforms like Facebook, LinkedIn, and Reddit and engage in conversations to demonstrate your expertise.
- **Referrals:** Ask your current clients for referrals or recommendations, as word-of-mouth marketing can be a powerful way to grow your client base.

The Importance of Building a Portfolio and Networking

A portfolio demonstrates your video writing skills and expertise, helping potential clients assess your suitability for their projects. To create an effective portfolio, include a variety of your best work, ideally within your chosen niche. You can also create a personal website or use portfolio platforms like Contently or Journo Portfolio to showcase your video script and description samples.

Networking is vital for finding clients and building long-term relationships. Connect with other video creators, producers, and content managers through online forums, social media platforms, and professional networking sites like LinkedIn. By engaging with others in the industry, you'll not only discover potential clients but also learn valuable insights and stay up-to-date with industry trends.

Tips for Marketing Your Services Effectively

- **Develop a personal brand:** Differentiate yourself from other video writers by highlighting your unique style and expertise.
- **Utilize social media:** Share your work and engage with your target audience on platforms like Twitter, LinkedIn, and Facebook.
- **Offer value:** Share valuable video content, insights, and tips within your industry to attract potential clients.
- **Showcase your expertise:** Offer to speak at conferences or events and write guest articles to demonstrate your video writing skills and knowledge.

MAXIMIZING YOUR EARNINGS POTENTIAL

Strategies for Increasing Your Rates and Earning More

- **Specialize in a niche:** Video writers with specialized knowledge or expertise can often charge higher rates.

- **Upsell related services:** Offer additional services like video editing, voice-over narration, or music composition to increase your earnings.
- **Obtain retainer agreements:** Establish ongoing relationships with clients who require regular video content, providing a stable income source.
- **Package deals:** Offer a package of video writing services, such as scriptwriting, video descriptions, and social media captions, to increase your earnings and overall value.

How to Negotiate with Clients

- **Know your worth:** Research industry rates and understand the value of your skills and expertise.
- **Be confident:** Clearly communicate your value proposition and justify your rates with examples of your work and results.
- **Offer tiered pricing:** Provide clients with multiple pricing options, allowing them to choose the level of service that fits their needs.
- **Be flexible:** Be open to negotiation but know your limits, and be prepared to walk away from

a project if the terms don't meet your requirements.

The Importance of Setting Goals and Tracking Your Progress

Setting clear goals helps you stay focused and motivated in your video writing business. Establish both short-term and long-term objectives, such as monthly income targets or the number of clients you want to acquire. Regularly track your progress and adjust your strategies as needed to stay on track toward your goals.

By following these steps and leveraging ChatGPT's capabilities, you can successfully establish a profitable side hustle in video script and description writing. Stay focused on your goals, continuously learn and improve, and adapt to the ever-changing digital landscape to ensure long-term success.

As you progress in your ChatGPT-assisted video script and description writing side hustle, always prioritize client satisfaction and maintain the highest quality of work. By consistently delivering top-notch content and nurturing strong relationships with your clients, you will not only secure repeat business but also establish a positive reputation within the industry.

3

COPYWRITING SERVICES WITH CHATGPT

Copywriting with ChatGPT is a side hustle where you use the advanced language capabilities of ChatGPT to create persuasive, engaging, and effective marketing copy for various platforms, such as websites, advertisements, social media posts, and email campaigns.

As a side hustler leveraging ChatGPT for copywriting services, you can enhance your productivity and expand your offerings, allowing you to cater to a wider range of clients in different industries.

USING CHATGPT FOR COPYWRITING: WRITING CONTENT FOR YOUR CLIENTS

Copywriting is the act of writing text, also known as copy, for the purpose of advertising or marketing a product, service, or idea. The goal of copywriting is to persuade the reader to take a specific action, such as buying a product, subscribing to a service, or signing up for a newsletter.

ChatGPT can be used to write content for your clients in various ways. Here are a few examples:

- **Generating headlines and taglines:** Headlines and taglines are crucial components of effective advertising and marketing. With ChatGPT, you can input keywords or phrases related to your client's business or product, and it will generate headlines and taglines that you can use in your copy.
- **Writing social media posts:** Social media is a powerful marketing tool, but creating content for social media can be time-consuming. With ChatGPT, you can input prompts and questions related to your client's social media goals, and it will generate content that you can use in your social media posts.

- **Crafting email campaigns:** Email marketing is an effective way to reach your client's target audience. With ChatGPT, you can input prompts related to your client's email campaign goals, and it will generate content for your emails.
- **Creating landing pages:** Landing pages are designed to convert website visitors into leads or customers. With ChatGPT, you can input prompts related to your client's landing page goals, and it will generate content for your landing page.

While ChatGPT can generate content for your clients, it's crucial to customize the content to match your client's brand voice and tone. It's also vital to ensure that the content is effective by testing and refining it as needed.

MAKING MONEY AS A CHATGPT COPYWRITER

How Much Can You Earn as a Copywriter?

As a freelance copywriter or someone pursuing copy-writing as a side hustle, your earnings can vary significantly based on your experience, niche, the type of clients you work with, and how much work you take on. Here's a general overview of potential earnings for a freelance or side hustle copywriter.

1. **Beginner copywriter:** As a beginner, you may start by charging lower rates while you build your portfolio and gain experience. Depending on the project, you can expect to earn anywhere from $25-$50 per hour or a few cents to $0.10 per word.
2. **Intermediate copywriter:** With some experience under your belt, you can raise your rates and attract higher-paying clients. At this stage, you might charge between $50-$100 per hour or $0.10-$0.50 per word.
3. **Experienced copywriter:** As an experienced freelance copywriter, you can charge premium rates typically ranging from $100-$150 or more per hour or $0.50-$1 or more per word.

Keep in mind that these figures are approximate. Your actual earnings can vary depending on the specific circumstances. As a freelancer or side hustler, your income will largely depend on the number of clients you have, your rates, and how much work you take on. To increase your earnings, consider specializing in a profitable niche, networking to find new clients, and consistently delivering high-quality work to build a strong reputation.

To make money as a ChatGPT copywriter, follow these steps.

- **Identify your niche:** Choose a specific industry or topic where you have knowledge and interest that will help you stand out in the market and make it easier to find clients who require specialized copywriting services.
- **Build a portfolio:** Create a portfolio showcasing your work, including marketing copy you have written with the help of ChatGPT. These samples will give potential clients an idea of your writing style and capabilities.
- **Set your rates:** Determine your pricing structure based on the value you provide, the time spent on each project, and the market rate for similar services.
- **Market your services:** Promote your copywriting services through social media, content platforms, and freelance job boards. Networking with others in the industry can also help you find potential clients.
- **Creating a unique selling proposition:** To stand out in a competitive market, it is crucial

to create a unique selling proposition (USP) that highlights the benefits of your ChatGPT copywriting services. These benefits can be based on your expertise, experience, turnaround time, or pricing structure.

- **Incorporating value-added services:** To differentiate your ChatGPT copywriting services from others in the market, consider offering value-added services such as SEO optimization, content strategy consulting, or social media management. These add-ons can help you attract clients who are looking for a full-service provider.

CHATGPT IN ACTION

Use the following prompt to generate email subject lines:

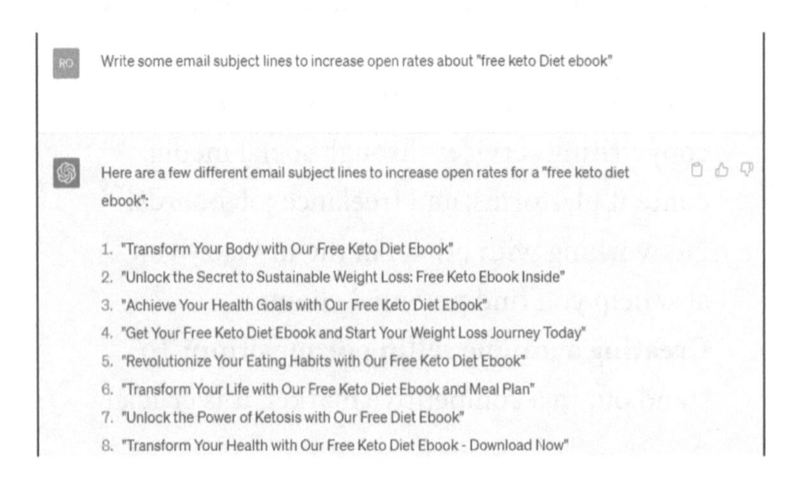

Use the following prompt to generate business taglines:

PROMPTS FOR COPYWRITING SERVICE USING CHATGPT

Headlines and Taglines

- Generate headlines and taglines for a new product launch.
- Come up with a catchy tagline for a business slogan.
- Create a compelling headline for a blog post.

Social Media Posts

- Write a social media post promoting a sale or promotion.

- Generate social media content for a product launch.
- Craft a social media post to engage with your client's followers.

Email Campaigns

- Write an email subject line to increase open rates.
- Create an email campaign to promote a new product.
- Generate content for a welcome email for new subscribers.

Landing Pages

- Write copy for a landing page to promote a service.
- Create content for a landing page to drive conversions.
- Generate copy for a landing page to promote a limited-time offer.

FINDING CLIENTS

To find clients for your ChatGPT-assisted copywriting services, consider these strategies.

- **Cold pitching:** Research businesses or individuals in your niche and send them a personalized email outlining your services and how you can help them improve their marketing efforts.
- **Job boards:** Regularly browse and apply for copywriting gigs on job boards like Upwork, Freelancer, and Fiverr.
- **Social media:** Join niche-specific groups and forums on platforms like Facebook, LinkedIn, and Reddit and engage in conversations to showcase your expertise.
- **Referrals:** Ask your current clients for referrals or recommendations, as word-of-mouth marketing can be a powerful way to grow your client base.
- **LinkedIn outreach:** Use LinkedIn to reach out to potential clients and offer your services. You can also use LinkedIn Sales Navigator to identify and target potential leads.
- **Content marketing:** Develop a robust content marketing strategy to showcase your expertise and attract potential clients. This strategy can include creating blog posts, social media content, or videos.
- **Collaborations:** Partner with other professionals or businesses in your niche to

offer joint services or collaborations. These partnerships can help you expand your reach and attract new clients.

- **Paid advertising:** Consider investing in paid advertising on platforms like Google Ads or Facebook Ads to reach a larger audience and attract potential clients.

TIPS FOR SUCCESS

1. **Understand marketing principles:** Familiarize yourself with the fundamentals of marketing, consumer psychology, and persuasion techniques to create compelling copy that drives results.
2. **Adapt to client's brand voice:** Learn to write in different styles and tones to match the unique brand voice of each client.
3. **Stay updated on industry trends:** Keep up with the latest trends and best practices in copywriting and digital marketing to ensure your copy remains relevant.
4. **Collaborate with clients:** Work closely with clients to understand their goals and target audience, ensuring that your copy aligns with their marketing objectives.

5. **Stay updated on ChatGPT advancements:** Regularly check for updates and improvements to the ChatGPT model, as they may enhance your productivity and the quality of your work.
6. **Diversify your services:** Offer additional services related to copywriting, such as content strategy consulting or social media management, to increase your income potential.

By following these steps and leveraging ChatGPT's capabilities, you can establish a profitable side hustle in copywriting services. Stay focused on your goals, continuously learn and improve, and adapt to the ever-changing digital landscape to ensure long-term success.

As you progress in your ChatGPT-assisted copywriting side hustle, always remember to prioritize client satisfaction and maintain the highest quality of work. By consistently delivering top-notch content and nurturing strong relationships with your clients, you will not only secure repeat business but also establish a positive reputation within the industry.

4

SOCIAL MEDIA MANAGEMENT AS A SIDE HUSTLE USING CHATGPT

In today's digital age, social media management has become an essential tool for businesses of all sizes to connect with their target audience, build brand awareness, and increase sales. As a result, there is a growing demand for skilled social media managers who can help businesses navigate the ever-changing landscape of social media platforms.

In this chapter, we will explore social media management as a lucrative side hustle and how ChatGPT can assist you in content creation, scheduling, finding clients, and maximizing your earnings potential.

SOCIAL MEDIA MANAGEMENT AS A SIDE HUSTLE

Social media management involves creating, scheduling, analyzing, and engaging with content posted on social media platforms, such as Facebook, Instagram, Twitter, LinkedIn, and Pinterest. As a social media manager, you will be responsible for:

- Developing and implementing a social media strategy.
- Creating engaging content for various platforms.
- Scheduling posts and managing content calendars.
- Monitoring and responding to comments and messages.
- Analyzing metrics and adjusting the strategy accordingly.

This side hustle is well-suited for individuals with strong communication skills, creativity, and an understanding of social media trends and platform algorithms. It also offers a flexible schedule and the ability to work remotely, making it an ideal side hustle for many people.

How Much Can You Earn with Social Media as a Side Hustle?

As a social media freelancer or side hustler, your earnings will hinge on your experience, niche, services offered, and the number of clients you work with.

In general, you can expect to earn anywhere from $30-$75 per hour or more for social media management, content creation, and strategy services. Keep in mind that these are approximate figures. Your actual earnings may vary based on the specific circumstances and the value you provide to your clients.

To increase your earnings, consider specializing in a specific platform or industry, building a strong portfolio, and networking to find new clients.

HOW CHATGPT CAN ASSIST WITH CONTENT CREATION AND SCHEDULING

ChatGPT, a powerful language model developed by OpenAI, can be a valuable tool for social media managers. It can help you with:

- **Generating content ideas:** ChatGPT can provide suggestions for blog posts, social media captions, and other types of content tailored to your client's niche and target audience.

- **Drafting content:** Use ChatGPT to draft engaging social media posts, saving you time and effort in the content creation process.
- **Proofreading and editing:** ChatGPT can help you identify grammar and punctuation errors, ensuring your content is polished and professional.
- **Scheduling posts:** Integrate ChatGPT with social media scheduling tools, like Buffer or Hootsuite, to automate the posting process, allowing you to focus on other tasks.

CHATGPT IN ACTION

Use the following prompt to create a social media content calendar:

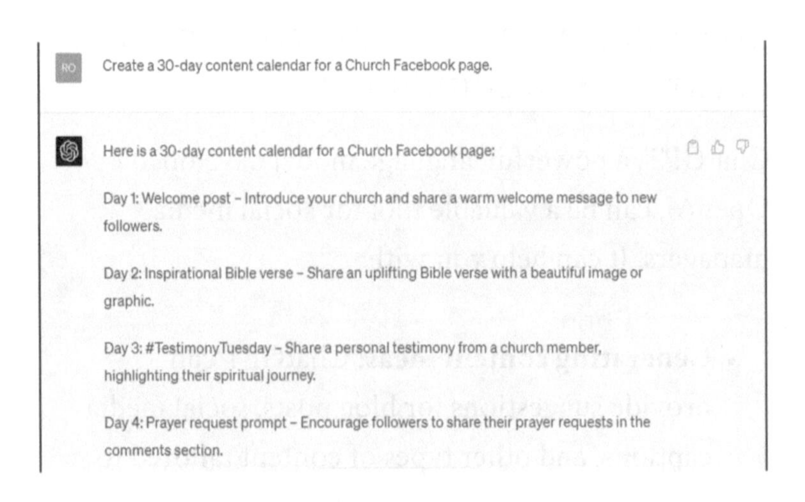

Use the following prompt to create a social media content calendar:

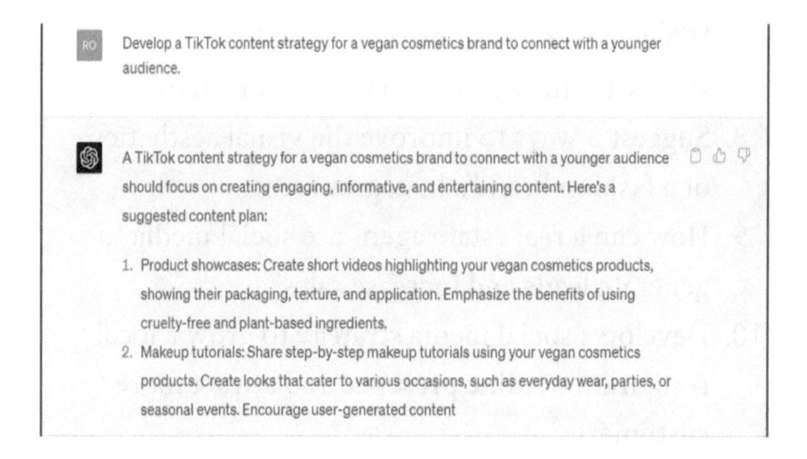

PROMPTS FOR SOCIAL MEDIA MANAGEMENT SIDE HUSTLE

1. Generate 10 engaging Instagram caption ideas for a fitness brand.
2. What are the top 5 social media trends for businesses in 2023?
3. Create a 30-day content calendar for a small bakery's Facebook page.
4. What are the best practices for increasing engagement on LinkedIn?
5. How can a travel agency effectively use Pinterest to attract more clients?

6. Draft a series of 5 tweets for a non-profit organization promoting their upcoming charity event.
7. What are the key metrics to track for a successful Instagram marketing campaign?
8. Suggest 3 ways to improve the visual aesthetics of a fashion brand's Instagram feed.
9. How can a real estate agent use social media to generate leads and increase sales?
10. Develop a social media strategy to grow a local restaurant's online presence and attract more customers.
11. What are the most effective techniques for growing a YouTube channel for a tech review business?
12. Generate a list of 10 Facebook post ideas for a veterinary clinic.
13. How can a personal finance blog leverage Twitter to increase their audience engagement?
14. What are the key elements of a successful influencer marketing campaign on Instagram?
15. Develop a TikTok content strategy for an eco-friendly cosmetics brand to connect with a younger audience.
16. Suggest 5 ways to optimize a LinkedIn company page for better visibility and engagement.

17. How can a podcast use social media platforms to increase its listenership?
18. Create a list of 10 Instagram Story ideas for a motivational speaker to inspire their followers.
19. What are the best practices for handling negative comments and feedback on social media?
20. How can a local bookstore use social media to create a sense of community and drive in-store traffic?

HOW TO FIND CLIENTS AND MARKET YOUR SERVICES

Building a client base is crucial for a successful social media management side hustle. Here are some strategies to help you find clients and market your services.

- **Develop a portfolio:** Create a portfolio showcasing your social media management skills, including examples of content you've created, accounts you've managed, and any relevant analytics.
- **Network:** Attend local business events, join online forums and social media groups related to your niche, and connect with potential clients on LinkedIn.

- **Offer free or discounted services:** Provide your services to friends, family, or local businesses for free or at a discounted rate to build your portfolio and gain testimonials.
- **Use freelance platforms:** Sign up for freelance websites, like Upwork, Fiverr, or Freelancer, to find clients looking for social media management services.
- **Promote your services on social media:** Create social media profiles for your side hustle and engage with potential clients through regular posting and interaction.

TIPS FOR MAXIMIZING YOUR EARNINGS POTENTIAL

To make the most of your social media management side hustle, follow these tips.

- **Upskill:** Stay current on social media trends and platform updates. Invest in learning new skills such as graphic design, photography, or copywriting to enhance your services.
- **Offer packages:** Create tiered service packages to cater to clients with different needs and budgets. Upsell additional services such as

content creation, analytics reporting, or ad management.

- **Specialize in a niche:** Focusing on a specific industry or niche can help you stand out from the competition and attract higher-paying clients.
- **Set clear expectations:** Establish clear expectations with clients regarding the scope of work, timelines, and deliverables to ensure smooth collaboration and minimize misunderstandings.
- **Be responsive:** Provide excellent customer service by promptly responding to client inquiries and addressing any concerns or issues that may arise.
- **Ask for referrals:** Encourage satisfied clients to refer your services to their network and consider offering incentives such as discounts or free services for successful referrals.
- **Optimize your workflow:** Streamline your processes and use tools like ChatGPT and social media scheduling platforms to increase efficiency, allowing you to take on more clients and grow your income.

Social media management is a promising side hustle with the potential for significant income and flexibility. By leveraging ChatGPT for content creation and scheduling, employing effective strategies to find clients and market your services, and following the tips provided to maximize your earnings, you can establish a successful social media management side hustle and enjoy the benefits of working in the rapidly evolving digital landscape.

5

WRITE RESUMES AND BIOS ON FIVERR AS A SIDE HUSTLE USING CHATGPT

In today's competitive job market, a well-crafted resume and professional bio can make all the difference for job seekers and professionals looking to advance their careers. As a result, there is a significant demand for skilled resume and bio writers who can help individuals present their skills, experiences, and achievements in the best possible light.

In this chapter, we will explore writing resumes and bios on Fiverr as a side hustle, how ChatGPT can assist you in this process, finding clients, and maximizing your earnings potential.

WRITING RESUMES AND BIOS ON FIVERR AS A SIDE HUSTLE

Fiverr is a popular online marketplace for freelance services, where individuals can offer their skills in various fields, including resume and bio writing. As a resume and bio writer, you will be responsible for:

- Understanding the client's background, skills, and goals.
- Crafting well-organized and visually appealing resumes.
- Writing engaging professional bios for various platforms.
- Tailoring content to specific industries and job positions.
- Proofreading and editing to ensure error-free documents.

This side hustle is ideal for individuals with strong writing, communication, and organizational skills, as well as an understanding of different industries and job requirements.

How Much Can You Earn as a Resume and Bio Writer on Fiverr?

As a side hustler writing resumes and bios on Fiverr, earnings depend on experience, quality, and client volume. Beginners may charge $5-$25 per project, intermediate writers $25-$75, and experienced writers $75-$200 or more.

Keep in mind Fiverr's 20% commission. To increase earnings, build a strong portfolio, gather positive reviews, and promote your services.

HOW CHATGPT CAN ASSIST WITH WRITING RESUMES AND BIOS

ChatGPT, a powerful language model developed by OpenAI, can be a valuable tool for resume and bio writers. It can help you with:

- **Generating content:** ChatGPT can provide suggestions for crafting compelling resume sections and professional bios, ensuring that the content is tailored to the client's needs and industry.
- **Drafting and editing:** Use ChatGPT to draft and edit resumes and bios, saving you time and effort in the writing process.

- **Proofreading:** ChatGPT can help you identify grammar and punctuation errors, ensuring that your documents are polished and professional.
- **Staying up-to-date:** ChatGPT can provide insights on current resume trends and best practices, enabling you to create resumes and bios that stand out in the job market.

CHATGPT IN ACTION

Use the following prompt to write a personal bio for a website:

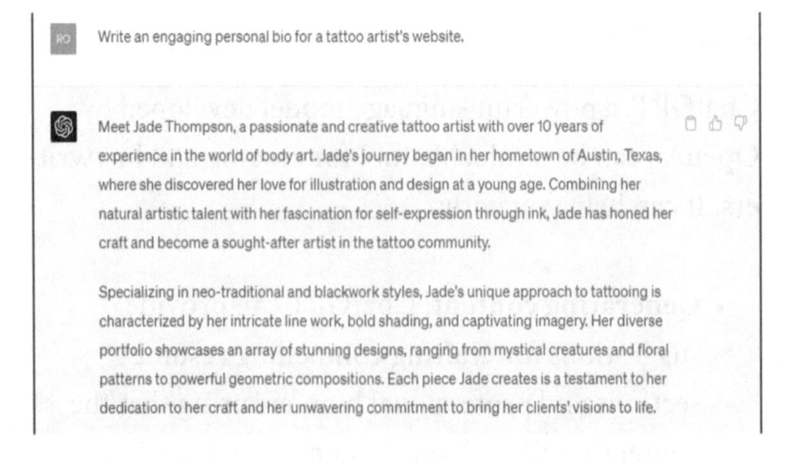

Use the following prompt to write a personal bio for a website:

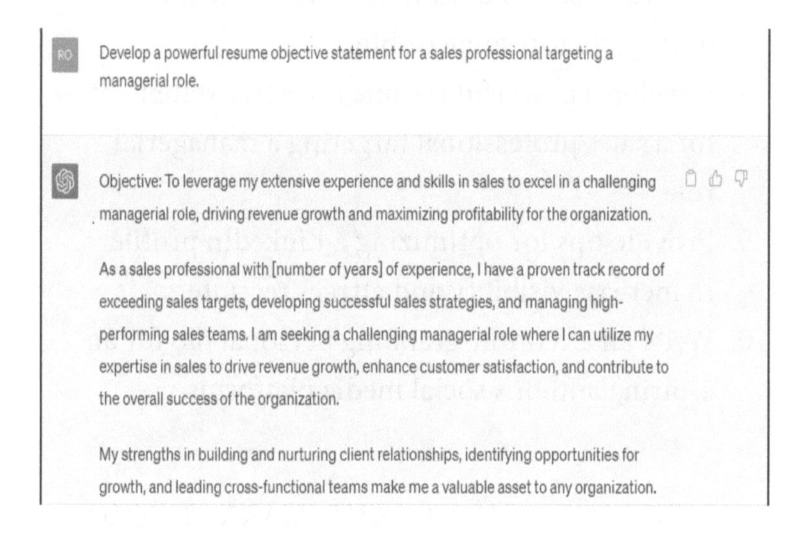
RO Develop a powerful resume objective statement for a sales professional targeting a managerial role.

Objective: To leverage my extensive experience and skills in sales to excel in a challenging managerial role, driving revenue growth and maximizing profitability for the organization.

As a sales professional with [number of years] of experience, I have a proven track record of exceeding sales targets, developing successful sales strategies, and managing high-performing sales teams. I am seeking a challenging managerial role where I can utilize my expertise in sales to drive revenue growth, enhance customer satisfaction, and contribute to the overall success of the organization.

My strengths in building and nurturing client relationships, identifying opportunities for growth, and leading cross-functional teams make me a valuable asset to any organization.

PROMPTS FOR WRITING RESUMES AND BIOS

1. Generate a compelling professional summary for a marketing manager with 5 years of experience.
2. Create a list of 10 action verbs to effectively describe accomplishments in a resume.
3. Draft a LinkedIn bio for a recent college graduate with a degree in computer science.
4. Suggest 5 resume formatting tips to create a visually appealing and easy-to-read document.
5. Write an engaging personal bio for a freelance graphic designer's website.

6. How can transferable skills be effectively showcased in a career change resume?
7. Create a list of 5 questions to ask a client before writing their resume or bio.
8. Develop a powerful resume objective statement for a sales professional targeting a managerial role.
9. Provide tips for optimizing a LinkedIn profile to increase visibility and attract recruiters.
10. Write an attention-grabbing personal bio for an aspiring author's social media platforms.

HOW TO FIND CLIENTS AND MARKET YOUR SERVICES

Building a client base is crucial for the success of your resume and bio writing side hustle on Fiverr. Here are some strategies to help you find clients and market your services.

- **Create an eye-catching gig:** Develop a Fiverr gig that showcases your resume and bio writing services, highlighting your expertise, experience, and unique selling points.
- **Build a portfolio:** Create a portfolio displaying your writing skills, including examples of

resumes and bios you've crafted for different industries and job positions.
- **Collect testimonials:** Request reviews and testimonials from satisfied clients to build credibility and trust with potential clients.
- **Promote your gig:** Share your Fiverr gig on social media platforms, LinkedIn, and relevant online forums to reach a wider audience.
- **Offer discounts or promotions:** Encourage new clients to try your services by offering limited-time discounts or promotional offers.

TIPS FOR MAXIMIZING YOUR EARNINGS POTENTIAL

To make the most of your resume and bio writing side hustle on Fiverr, follow these tips.

- **Upskill:** Stay current on resume trends and best practices. Invest in learning new skills related to resume and bio writing, such as graphic design or search engine optimization (SEO).
- **Offer add-on services:** Provide additional services such as cover letter writing, LinkedIn profile optimization, or career coaching to increase your earnings potential.

- **Specialize in a niche:** Focusing on a specific industry or job function can help you stand out from the competition and attract higher-paying clients.
- **Maintain high-quality standards:** Ensure that your work is consistently high quality, meeting or exceeding client expectations to garner positive reviews and repeat business.
- **Be responsive:** Provide excellent customer service by promptly responding to client inquiries and addressing any concerns or issues that may arise.
- **Optimize your gig:** Regularly update and optimize your Fiverr gig, using relevant keywords and eye-catching images to improve its visibility and attract more clients.
- **Set clear expectations:** Establish clear expectations with clients regarding the scope of work, timelines, and deliverables to ensure smooth collaboration and minimize misunderstandings.
- **Manage your time efficiently:** Streamline your processes and use tools like ChatGPT to increase efficiency, allowing you to take on more clients and grow your income.

Writing resumes and bios on Fiverr is a promising side hustle with the potential for significant income and flexibility. By harnessing ChatGPT for content creation and editing, employing effective strategies to find clients and market your services, and following the tips provided to maximize your earnings, you can establish a successful resume and bio writing side hustle and enjoy the benefits of working in the ever-evolving freelance marketplace.

Let Other Gig Workers Know About 10 Side Hustles that Can Earn Them Extra Income With the Help of ChatGPT!

> *"Once a new technology rolls over you, if you're not part of the steamroller, you're part of the road."*
>
> — *STEWART BRAND, WRITER*

Remember in the introduction, when we mentioned that millions of people are making an extra income by running a side hustle or part-time venture?

The gig economy currently accounts for a third of the world's working population and is projected to reach a worth of $500 billion over the next five years.

Cloud sharing, software like Slack and Asana, and video conferencing tools like Zoom have all done their share to streamline remote work and empower gig workers to pursue success in their own time, wherever they are located.

Without a doubt, one of the most groundbreaking gig economy solutions in current times is ChatGPT. This AI-driven language tool reached one million users within five days of its launch and 100 million within two months!

By this point in this book, you have seen how ChatGPT is an infallible tool for writing blogs and website content, creating video scripts and descriptions, copywriting, social media management, and resume and bio creation.

I have also provided you with powerful strategies for finding clients in key sectors and maximizing your earnings.

So many successful businesses use ChatGPT on a daily basis for everything from customer service to sales and marketing. Salesforce, Slack, and Duolingo are just a few organizations that are improving their products and services with this revolutionary technology.

Many freelancers are following their lead, making their tasks easier quicker, and more time-efficient than ever before—and sharing their discoveries with other ChatGPT users.

And you can do the same. You can play a big role in helping others gain control over their finances and increase their monthly income by double or more, by letting them know about ChatGPT's wide range of abilities.

This is *your* chance to give other freelance workers the guidance they need.

By leaving a review of this book on Amazon, you can share key prompts that others can use to obtain the kind of answers from ChatGPT that can result in impressive earnings.

When you leave your honest opinion of this book and how it's helped you on Amazon, other readers will understand the myriad of ways in which ChatGPT can help them build a better future for themselves and their families.

Thank you for helping people like you find the guidance they're looking for. ChatGPT could be one of the most efficient "personal assistants" you will encounter, but sharing points of view is still the exclusive domain of human beings.

6

VIRTUAL ASSISTANCE AS A SIDE HUSTLE USING CHATGPT

The growing trend of remote work and the need for businesses to manage various tasks has driven the demand for virtual assistants. As a virtual assistant, you can offer a wide range of administrative and support services to clients from the comfort of your home.

In this chapter, we will explore virtual assistance as a side hustle and how ChatGPT can assist you with administrative tasks, finding clients, and maximizing your earnings potential.

VIRTUAL ASSISTANCE AS A SIDE HUSTLE

A virtual assistant (VA) is a self-employed professional who provides remote administrative support to clients, such as entrepreneurs, small business owners, or busy professionals. Some common tasks a VA might handle include:

- Calendar and email management
- Data entry and organization
- Social media management
- Customer service
- Research and content creation

This side hustle is ideal for individuals with strong organizational and communication skills, as well as the ability to work independently and manage their time effectively.

How Much Can You Earn as a Virtual Assistant?

A virtual assistant's earnings can vary depending on experience, location, skills, and the type of clients they work with. Generally, virtual assistants can expect to earn anywhere from $10-$50 per hour or more.

Here's a rough breakdown:

1. **Entry-level virtual assistant:** $10-$20 per hour.
2. **Intermediate-level virtual assistant:** $20-$35 per hour.

3. **Experienced virtual assistant:** $35-$50+ per hour.

These figures are approximate, and actual earnings may vary. To increase your earnings as a virtual assistant, focus on developing specialized skills, gaining experience in a specific niche or industry, and building a strong reputation with clients through consistent, high-quality work.

HOW CHATGPT CAN ASSIST WITH ADMINISTRATIVE TASKS

ChatGPT, a powerful language model developed by OpenAI, can be a valuable tool for virtual assistants. It can help you with:

- **Drafting and editing emails:** ChatGPT can help you draft, edit, and proofread emails for clients, ensuring professional and polished communication.
- **Content creation:** Generate engaging content for blog posts, social media updates, and other client needs with the help of ChatGPT.
- **Research:** Use ChatGPT to gather information on various topics, helping you save time and provide accurate data to your clients.

- **Note-taking and summarizing:** ChatGPT can help you take notes and create summaries from meetings or documents, providing clients with concise and organized information.
- **Task management:** Get assistance in organizing tasks, setting priorities, and creating to-do lists to stay on top of your workload.

CHATGPT IN ACTION

Use the following prompt to write a professional website for a virtual assistant:

RO

Describe the process of setting up a professional virtual assistant website to attract potential clients.

Setting up a professional virtual assistant website can be a great way to showcase your skills and attract potential clients. Here is a step-by-step process to follow:

1. Define your target audience: Identify the type of clients you want to attract and tailor your website accordingly. This will help you create content that resonates with your ideal clients.
2. Choose a domain name: Choose a domain name that is easy to remember, relevant to your business, and available. Consider using your name or keywords related to your services.

Use the following prompt for a virtual assistant:

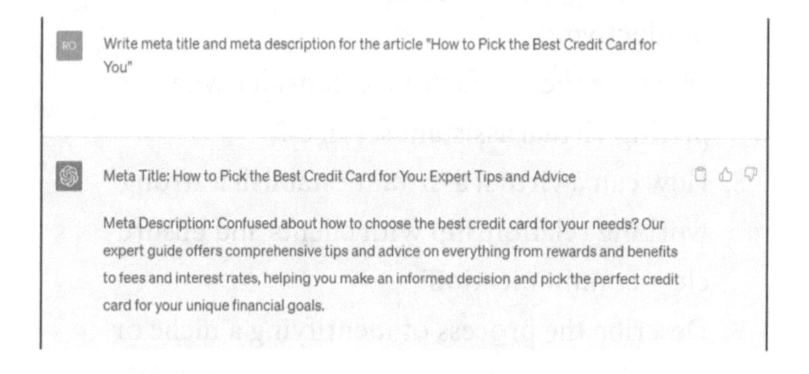

PROMPTS FOR VIRTUAL ASSISTANCE AS A SIDE HUSTLE

1. List 10 essential skills that every successful virtual assistant should possess.
2. What are the most effective strategies for managing multiple clients and tasks as a virtual assistant?
3. Describe the process of setting up a professional virtual assistant website to attract potential clients.
4. How can a virtual assistant use social media to expand their network and market their services effectively?
5. Discuss the benefits and challenges of working as a virtual assistant from a remote location.

6. Provide a list of 5 tools and apps that can help virtual assistants stay organized and productive.
7. What are the key factors to consider when pricing virtual assistant services?
8. How can a virtual assistant establish a strong working relationship with clients and ensure clear communication?
9. Describe the process of identifying a niche or specialty as a virtual assistant and its potential impact on earnings.
10. Outline the steps to create an effective marketing plan for a virtual assistant side hustle.

HOW TO FIND CLIENTS AND MARKET YOUR SERVICES

Building a client base is essential for the success of your virtual assistant side hustle. Here are some strategies to help you find clients and market your services.

- **Create a professional website:** Develop a website highlighting your virtual assistance services, skills, experience, and the range of tasks you can handle.

- **Network on social media and LinkedIn:** Connect with potential clients and engage in relevant groups and discussions to showcase your expertise and attract clients.
- **Register on freelance platforms:** Sign up for platforms like Upwork, Freelancer, or Belay, which connect virtual assistants with clients.
- **Offer a free trial or discounted services:** Encourage new clients to try your services by offering a limited-time free trial or discounted rates.
- **Request referrals and testimonials:** Ask satisfied clients to refer you to their networks and provide testimonials to build credibility and trust.

TIPS FOR MAXIMIZING YOUR EARNINGS POTENTIAL

To make the most of your virtual assistant side hustle, follow these tips.

- **Specialize in a niche:** Focusing on a specific industry or area of expertise can help you stand out from the competition and attract higher-paying clients.

- **Upskill:** Learn new skills, such as graphic design, project management, or social media management, to expand your service offerings and increase your income potential.
- **Set clear expectations:** Establish clear expectations with clients regarding the scope of work, timelines, and deliverables to ensure smooth collaboration and minimize misunderstandings.
- **Be responsive:** Provide excellent customer service by promptly responding to client inquiries and addressing any concerns or issues that may arise.
- **Manage your time effectively:** Use time management techniques and tools like ChatGPT to increase efficiency, allowing you to take on more clients and grow your income.

Virtual assistance can be a good side job that pays well and offers flexibility. You can use ChatGPT to help with administrative tasks, find clients by promoting your services, and earn more money by following the tips given.

It's important to offer different services, keep learning about the job, and improve your skills to stay ahead. With a successful virtual assistant business, you can work from home and be part of the gig economy.

7

WRITING AND SELLING CHILDREN'S BOOKS AS A SIDE HUSTLE USING CHATGPT

The children's book market is a vibrant and growing industry, allowing creative individuals to share their stories with young readers while generating income.

In this chapter, we will explore writing and selling children's books as a side hustle, how ChatGPT can assist you in the writing and research process, the steps to publish and market your books, and tips for maximizing your earnings potential.

WRITING AND SELLING CHILDREN'S BOOKS AS A SIDE HUSTLE

Writing children's books as a side hustle is ideal for those who have a knack for storytelling, a vivid imagination, and a passion for inspiring young readers. As a children's book author, you will be responsible for:

- Developing engaging story ideas.
- Creating memorable characters.
- Writing age-appropriate and captivating narratives.
- Collaborating with illustrators (if necessary).
- Publishing and marketing your books.

This side hustle can be both rewarding and profitable, as successful children's books have the potential for high sales and ongoing royalties.

How Much Can You Earn by Writing and Selling Children's Books?

Based on the provided information, earnings from writing and selling children's books can be categorized into three levels:

1. **Most children's book authors:** These authors typically earn around $20,000 per year, reflecting the majority of writers in the field.
2. **Serious authors:** More involved and dedicated authors can expect to earn between $30,000 and $70,000 per year as they invest more time and effort into their craft.

3. **Top 1% of children's book authors:** The most successful children's book authors can make over $200,000 per year, reflecting their exceptional success in the market.

It's important to note that earnings from writing and selling children's books can be unpredictable and vary greatly among authors. To increase your chances of success, focus on writing engaging stories and investing time in marketing and promotion to reach your target audience.

HOW CHATGPT CAN ASSIST WITH WRITING AND RESEARCH

ChatGPT, a powerful language model developed by OpenAI, can be a valuable tool for aspiring children's book authors. It can help you with:

- **Brainstorming story ideas:** Generate unique and engaging story ideas for various age groups and genres.
- **Developing characters:** ChatGPT can help you create memorable and relatable characters that resonate with young readers.
- **Drafting and editing:** Use ChatGPT to draft and edit your stories, ensuring that the

language is age-appropriate and the narrative is engaging.

- **Research:** ChatGPT can gather information on children's book trends, popular themes, and best practices to ensure your books resonate with your target audience.
- **Feedback and suggestions:** Use ChatGPT to get feedback on your story drafts and suggestions for improvement.

CHATGPT IN ACTION

Use the following prompt to generate story ideas for children's books:

PROMPTS FOR WRITING AND SELLING CHILDREN'S BOOKS

1. What are the key differences between traditional publishing and self-publishing children's books, and which route is best for a side hustle?
2. How can an aspiring children's book author determine the appropriate age group and reading level for their story?
3. List 10 popular themes and genres in children's literature to inspire story ideas.
4. Discuss the process of finding and collaborating with an illustrator for your children's book.
5. What are the essential elements of a successful children's book marketing campaign?
6. Explain the role of literary agents in the traditional publishing process and how to find the right agent for your children's book.
7. Provide a step-by-step guide for self-publishing a children's book on Amazon Kindle Direct Publishing.
8. How can authors use social media and other online platforms to build a loyal readership and promote their children's books?

9. Discuss strategies for creating diverse and inclusive characters in children's literature.
10. What are the benefits and challenges of writing a series of children's books? How can these benefits and challenges impact earnings potential?

HOW TO PUBLISH AND MARKET YOUR BOOKS

To successfully publish and market your children's books, follow these steps.

- **Choose a publishing route:** Decide whether to pursue traditional publishing (submitting your manuscript to literary agents and publishers) or self-publishing (using platforms like Amazon Kindle Direct Publishing or IngramSpark).
- **Collaborate with an illustrator:** If your book requires illustrations, find a talented illustrator to bring your story to life. Platforms like Fiverr and Behance can help you connect with illustrators.
- **Edit and proofread:** Ensure that your manuscript is polished and error-free by working with a professional editor or using tools like Grammarly and ProWritingAid.

- **Design a captivating cover:** Create an eye-catching cover that appeals to your target audience, either by working with a graphic designer or using online design tools like Canva.
- **Market your book:** Promote your book through social media, email marketing, local events, and collaborations with influencers or other authors. Submit your book for reviews and consider running paid advertising campaigns.

TIPS FOR MAXIMIZING YOUR EARNINGS POTENTIAL

To make the most of your children's book side hustle, consider these tips.

- **Write a series:** Writing a series of books can help you build a loyal readership and increase your earnings potential as readers eagerly await each new installment.
- **Diversify your target age groups:** Write for different age groups to cater to a broader audience, such as picture books for preschoolers, early reader books, and middle-grade novels.

- **Explore various formats:** Experiment with different formats, such as ebooks, print books, and audiobooks, to reach a broader market.
- **Network with other authors and professionals:** Connect with other children's book authors, illustrators, and industry professionals to learn from their experiences and gain exposure to your work.
- **Attend conferences and workshops:** Participate in children's book conferences and workshops to hone your craft, network with industry professionals, and stay up-to-date with market trends and publishing opportunities.

Writing and selling children's books can be a fulfilling and profitable side hustle. You can successfully launch your children's book career by utilizing ChatGPT for writing and research assistance, selecting the appropriate publishing route, and employing effective marketing strategies.

To maximize your earning potential, diversify your target age groups, experiment with different formats, and establish a network within the industry.

As you continue to write and share your stories with young readers, your side hustle can grow into a thriving business. You have the potential to leave a lasting impression on the world of children's literature.

8

ONLINE SURVEY PARTICIPATION AS A SIDE HUSTLE USING CHATGPT

Online survey participation is a flexible and accessible side hustle that allows individuals to earn extra income by sharing their opinions and insights.

In this chapter, we will explore online survey participation as a side hustle, how to find survey opportunities and maximize earnings, how ChatGPT can improve survey responses and efficiency, and tips for maximizing your earnings potential.

ONLINE SURVEY PARTICIPATION AS A SIDE HUSTLE

Many market research companies, academic institutions, and businesses seek valuable consumer input through online surveys. As a survey participant, you can earn money by completing these surveys, which may cover various topics, such as:

- Product reviews and preferences
- Consumer habits and trends
- Political opinions
- Media and entertainment preferences
- Health and lifestyle choices

This side hustle is ideal for individuals who are looking for a low-commitment way to earn extra income during their free time.

How Much Can You Earn by Online Survey Participation as a Side Hustle?

Earnings from online survey participation as a side hustle can differ based on several factors, such as the platform used, the frequency and duration of the surveys, the target demographic, and the country in which you live. In general, online surveys are not the most lucrative side hustle, but they can still provide a modest source of extra income.

On average, you can expect to earn anywhere from a few cents to a couple of dollars per survey. The length and complexity of the survey, as well as the rarity of the target demographic, will influence the payment amount. Some survey sites also offer a points-based system that can be redeemed for cash or gift cards.

To maximize your earnings, it's a good idea to sign up for multiple survey platforms. Doing so will increase the number of available surveys and opportunities for income.

While participating in online surveys can be an easy way to make extra money, it is unlikely to provide a substantial income. Many people earn between $50-$200 per day from survey participation, but this income can fluctuate depending on the factors mentioned above.

HOW TO FIND SURVEY OPPORTUNITIES AND MAXIMIZE EARNINGS

To find online survey opportunities and maximize your earnings, follow these steps.

- **Sign up for multiple survey sites:** Register on various survey platforms, such as Swagbucks, Survey Junkie, or Vindale Research, to increase the number of survey invitations you receive.
- **Complete your profile:** Fill out your demographic and interest information as

thoroughly as possible to ensure you receive survey invitations tailored to your background and preferences.

- **Prioritize high-paying surveys:** Focus on completing surveys that offer higher payouts or rewards, as they can significantly increase your earnings potential.
- **Check your email frequently:** Survey invitations are often sent via email, so check your inbox regularly to avoid missing out on opportunities.
- **Be honest and consistent:** Provide honest and consistent responses to maintain a good reputation as a survey participant and receive more invitations.

HOW TO USE CHATGPT TO IMPROVE SURVEY RESPONSES AND EFFICIENCY

ChatGPT can be a valuable tool for online survey participants. It can help you with:

- **Developing coherent and well-structured responses:** ChatGPT can help you craft clear and concise answers to open-ended survey questions, ensuring your feedback is useful to researchers.

- **Generating ideas and opinions:** If you're unsure how to respond to a survey question, ChatGPT can provide suggestions and ideas to help you form an opinion.
- **Saving time:** Use ChatGPT to quickly generate answers to survey questions, allowing you to complete more surveys in less time and increase your earnings potential.

CHATGPT IN ACTION

Use the following prompt research on survey platforms:

RO How can you identify high-quality survey platforms and avoid scams when looking for online survey opportunities?

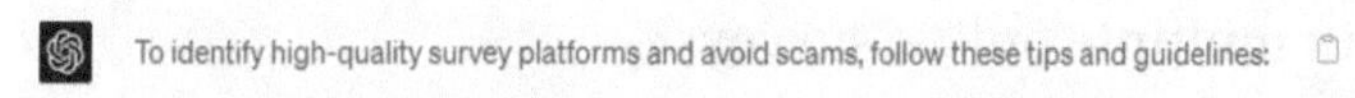
To identify high-quality survey platforms and avoid scams, follow these tips and guidelines:

1. Research the platform: Before signing up for a survey site, research its reputation and history. Check for reviews from reputable sources, user testimonials, and any red flags that may indicate a scam.
2. Look for transparency: Legitimate survey platforms should provide clear information about their company, privacy policy, terms and conditions, and how they handle your personal information. Be cautious if a site is vague or provides little information about these aspects.
3. Verify payment methods: High-quality survey platforms

PROMPTS FOR ONLINE SURVEY PARTICIPATION

1. How can you identify high-quality survey platforms and avoid scams when looking for online survey opportunities?
2. What strategies can help you stay motivated and consistent in your online survey participation side hustle?
3. Discuss the importance of honesty and consistency in your survey responses and how these qualities affect your reputation as a survey participant.
4. What are the best practices for managing your time and staying organized when participating in multiple online surveys?
5. How can you effectively handle open-ended survey questions to provide valuable and actionable feedback to researchers?
6. Describe the process of using ChatGPT to generate ideas and opinions when you're unsure how to respond to a survey question.
7. What are some common mistakes to avoid when participating in online surveys to ensure a positive experience and maximize earnings?
8. Discuss the role of demographic and interest information in receiving tailored survey

invitations and how to optimize your profile for better opportunities.

9. How can you balance online survey participation with other side hustles to diversify your income and achieve greater financial stability?
10. Provide tips on creating a comfortable and efficient workspace for online survey participation to maximize productivity and minimize distractions.

TIPS FOR MAXIMIZING YOUR EARNINGS POTENTIAL

To make the most of your online survey participation side hustle, consider these tips.

- **Set a daily or weekly goal:** Establish a routine by setting aside a specific amount of time each day or week for survey participation, helping you stay consistent and focused.
- **Stay organized:** Keep track of your survey earnings and invitations using a spreadsheet or dedicated email folder to maximize your income potential.
- **Look for referral bonuses:** Some survey platforms offer referral bonuses when you

invite friends or family members to join. Take advantage of these opportunities to boost your earnings.

- **Be patient and persistent:** Building up earnings from online surveys takes time and dedication. Stay patient and persistent, and you'll see your income grow over time.
- **Diversify your side hustle:** Combine online survey participation with other side hustles, such as freelance writing or virtual assistance, to increase your overall earnings potential.

Online survey participation can be a simple and flexible way to earn extra income in your spare time. By finding survey opportunities, maximizing earnings, and using ChatGPT to improve survey responses and efficiency, you can make the most of this side hustle.

With patience, persistence, and a commitment to providing valuable feedback, you can contribute to your earnings while helping businesses and researchers gather valuable insights.

Remember to stay organized, prioritize high-paying surveys, and set goals to maximize your earning potential. Furthermore, consider combining online survey participation with other side hustles to diversify your income streams. By following these tips and leveraging the power of ChatGPT, you can turn online survey participation into a successful and rewarding side hustle.

9

ONLINE CUSTOMER SUPPORT AS A SIDE HUSTLE USING CHATGPT

Providing online customer support is a popular side hustle for those who enjoy helping others and possess excellent communication skills.

In this chapter, we will explore online customer support as a side hustle, how ChatGPT can assist you in handling customer service tasks, how to find clients and market your services, and tips for maximizing your earnings potential.

ONLINE CUSTOMER SUPPORT AS A SIDE HUSTLE

As an online customer support representative, you will assist customers with their questions, concerns, and issues, typically via chat, email, or phone. Tasks may include:

- Resolving customer complaints.
- Providing product or service information.
- Processing orders, returns, or refunds.
- Troubleshooting technical issues.
- Giving recommendations and advice.

This side hustle is ideal for individuals who possess strong problem-solving skills, empathy, and patience.

How Much Can You Earn with Online Customer Support as a Side Hustle?

With an online customer support side hustle, you can expect to earn anywhere from $10-$35 per hour, depending on your experience, the complexity of the job, and the company you work for.

However, earnings can vary and may be higher for those with specialized skills or in certain industries.

HOW CHATGPT CAN ASSIST WITH CUSTOMER SERVICE TASKS

ChatGPT, a powerful language model developed by OpenAI, can be an invaluable tool for online customer support representatives. It can help you with:

- **Drafting professional responses:** Use ChatGPT to generate well-structured and empathetic responses to customer inquiries, ensuring a positive experience.
- **Troubleshooting assistance:** ChatGPT can help you find solutions to common technical issues or problems, enabling you to provide efficient support to customers.
- **Knowledge base:** Leverage ChatGPT to quickly access product or service information, policies, and guidelines to better assist customers.
- **Language assistance:** ChatGPT can help you communicate with customers who speak different languages, allowing you to support a more diverse clientele.
- **Time management:** ChatGPT can help you save time by providing instant, relevant information and suggestions, allowing you to handle more customer interactions and increase your earnings potential.

CHATGPT IN ACTION

Use the following prompt to get tips to improve customer support side hustle:

RO

How can you effectively manage your time and multitask when handling multiple customer inquiries simultaneously?

Effectively managing your time and multitasking when handling multiple customer inquiries simultaneously can be challenging but is achievable with practice and the right approach. Here are some tips to help you:

1. Prioritize tasks: Assess the urgency and importance of each inquiry to determine which tasks require immediate attention. Address high-priority tasks first and work your way down the list.

PROMPTS FOR ONLINE CUSTOMER SUPPORT AS A SIDE HUSTLE

1. What are the best practices for providing excellent online customer support in a chat-based environment?
2. How can you effectively manage your time and multitask when handling multiple customer inquiries simultaneously?
3. Discuss strategies for developing a personal and professional rapport with customers during online interactions.

4. What are the main differences between providing customer support via chat, email, and phone, and how can you adapt your communication style accordingly?
5. How can you effectively handle language barriers and cultural differences when providing online customer support to an international clientele?
6. Describe the process of tracking and measuring key performance indicators (KPIs) in online customer support to ensure consistent service quality.
7. What are some common customer support scenarios, and how can you prepare for them to provide quick assistance?
8. Discuss the importance of staying up-to-date with company policies, product updates, and industry news to provide accurate and relevant support to customers.
9. How can you manage and reduce stress in a high-pressure online customer support environment?
10. Provide tips for developing a growth mindset and continuously learning and improving as an online customer support representative.

HOW TO FIND CLIENTS AND MARKET YOUR SERVICES

To find clients and market your online customer support services, follow these steps.

- **Create a professional profile:** Develop a profile on freelancing platforms like Upwork, Freelancer, or PeoplePerHour, highlighting your customer support skills, experience, and availability.
- **Network:** Join online forums, social media groups, and LinkedIn communities related to customer support and your industry to connect with potential clients and fellow professionals.
- **Offer a niche service:** Specialize in a specific industry or type of customer support (e.g., technical support, e-commerce) to differentiate yourself from competitors and attract clients with specific needs.
- **Reach out to local businesses:** Offer your services to local businesses or startups that may require customer support but lack the resources for a full-time employee.
- **Showcase your expertise:** Write blog posts or create videos demonstrating your knowledge of customer support best practices, tools, and

strategies to establish credibility and attract clients.

TIPS FOR MAXIMIZING YOUR EARNINGS POTENTIAL

To make the most of your online customer support side hustle, consider these tips.

- **Develop specialized skills:** Invest in learning specialized skills, such as technical support, customer success, or CRM software expertise, to increase your value to clients and command higher rates.
- **Offer multiple communication channels:** Provide support via chat, email, and phone to cater to various customer preferences and increase your client base.
- **Set clear boundaries:** Establish your working hours and availability to maintain a healthy work-life balance while managing your side hustle.
- **Seek long-term clients:** Aim to build long-term relationships with clients, as consistent work can lead to increased earnings and stability in your side hustle.

- **Continuously improve:** Regularly seek feedback from clients and customers to identify areas for improvement and enhance your customer support skills.

Online customer support can be a rewarding and profitable side hustle for individuals with strong communication and problem-solving skills. This side hustle involves assisting customers with their inquiries and issues through chat, email, or phone.

ChatGPT can aid in drafting responses, troubleshooting, providing knowledge, language assistance, and time management. To find clients, create a professional profile, network, offer niche services, reach out to local businesses, and showcase your expertise.

To maximize earnings, develop specialized skills, offer multiple communication channels, set clear boundaries, seek long-term clients, and continuously improve.

10

LANGUAGE TRANSLATION SERVICES AS A SIDE HUSTLE USING CHATGPT

Language translation services are in high demand as businesses and individuals seek to bridge communication gaps in our increasingly globalized world.

In this chapter, we will explore language translation as a side hustle, how ChatGPT can assist with translation tasks, how to find clients and market your services, and tips for maximizing your earnings potential.

LANGUAGE TRANSLATION SERVICES AS A SIDE HUSTLE

As a freelance translator, you will convert written content from one language to another while maintaining the original meaning and context. Translation services are needed across various industries, including:

- Legal documents
- Medical records
- Marketing materials
- Technical manuals
- Websites and blogs

This side hustle is ideal for individuals who are fluent in two or more languages and possess strong written communication skills.

How Much Can You Earn from Language Translation Services?

Earnings from language translation services as a side hustle can vary significantly based on the language pair, industry specialization, level of expertise, and the complexity of the documents. Here is a general breakdown of potential earnings for different types of translations in a side hustle capacity.

1. **Legal documents:** Translating legal documents as a side hustle usually requires a high level of expertise and knowledge of legal terminology. You can expect to earn between $0.10-$0.25

per word, with the potential for higher earnings as you gain more experience.

2. **Medical records:** As a side hustle, medical translations require specialized knowledge and can be quite lucrative. Rates can range from $0.10-$0.25 per word, depending on the complexity of the document and your experience.
3. **Marketing materials:** Translating marketing materials in a side hustle capacity often requires creativity and an understanding of the target audience's culture. Rates for marketing translations can range from $0.08-$0.20 per word.
4. **Technical manuals:** Translating technical manuals as a side gig typically involves complex terminology and a strong understanding of the subject matter. Rates can vary from $0.10-$0.25 per word, depending on the industry and level of expertise required.
5. **Websites and blogs:** Translating web content as a side hustle can be less specialized but still requires strong language skills and cultural understanding. Rates for website and blog translations generally range from $0.05-$0.15 per word.

Keep in mind that these rates are approximate and can vary based on your location, experience, and specific project requirements. Additionally, some translators charge hourly rates or project-based fees instead of per-word rates. As you gain more experience and build a strong reputation in your work, you can potentially earn more for your translation services.

HOW CHATGPT CAN ASSIST WITH TRANSLATION TASKS

ChatGPT can be an invaluable tool for freelance translators. It can help you with:

- **Drafting translations:** Use ChatGPT to generate initial translations of text, which you can then review and refine to ensure accuracy and proper context.
- **Proofreading:** ChatGPT can help you identify grammatical errors, inconsistencies, or awkward phrasings in your translations, ensuring a polished final product.
- **Research assistance:** ChatGPT can provide context and background information on specific terms, phrases, or cultural references, allowing you to produce more accurate and appropriate translations.

- **Language assistance:** ChatGPT supports multiple languages, making it a useful tool for translators working with various language pairs.

CHATGPT IN ACTION

Use the following prompt to translate English text to Spanish:

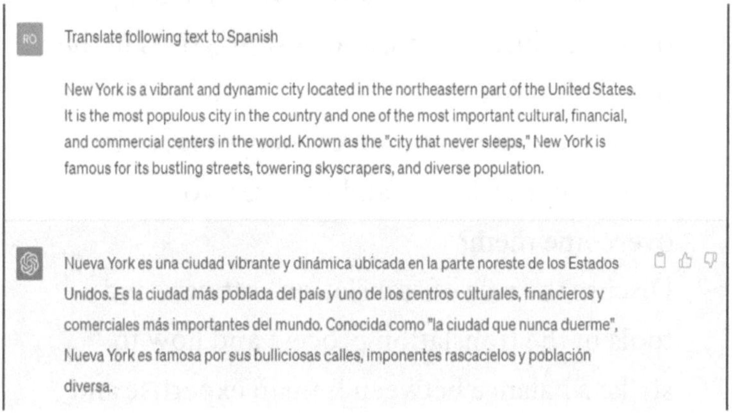

RO Translate following text to Spanish

New York is a vibrant and dynamic city located in the northeastern part of the United States. It is the most populous city in the country and one of the most important cultural, financial, and commercial centers in the world. Known as the "city that never sleeps," New York is famous for its bustling streets, towering skyscrapers, and diverse population.

Nueva York es una ciudad vibrante y dinámica ubicada en la parte noreste de los Estados Unidos. Es la ciudad más poblada del país y uno de los centros culturales, financieros y comerciales más importantes del mundo. Conocida como "la ciudad que nunca duerme", Nueva York es famosa por sus bulliciosas calles, imponentes rascacielos y población diversa.

PROMPTS FOR LANGUAGE TRANSLATION SERVICES

1. What are the key skills and qualities required for success in the language translation services industry?

2. Discuss the importance of cultural awareness and sensitivity when providing language translation services.
3. Explain the differences between translation, localization, and transcreation and how to determine which service is most appropriate for a given project.
4. Describe the process of creating a strong portfolio to showcase your translation skills and attract potential clients.
5. How can you stay up-to-date with language trends, industry news, and best practices in the translation field?
6. What are some common challenges faced by freelance translators, and how can you overcome them?
7. Discuss the role of translation software and tools in the translation process and how to strike a balance between human expertise and technological assistance.
8. How can you effectively manage your time and workload when juggling multiple translation projects and deadlines?
9. What strategies can help you build a strong network of clients and fellow translators to support your side hustle?

10. Provide tips for negotiating rates and contracts with clients to ensure fair compensation and clear expectations for your translation services.

HOW TO FIND CLIENTS AND MARKET YOUR SERVICES

To find clients and market your translation services, follow these steps.

- **Create a professional profile:** Develop a profile on freelancing platforms like Upwork, Freelancer, or ProZ, highlighting your language skills, translation experience, and areas of expertise.
- **Network:** Join online forums, social media groups, and LinkedIn communities related to translation and your language pairs to connect with potential clients and fellow professionals.
- **Offer a niche service:** Specialize in a specific industry or type of translation (e.g., legal, medical, technical) to differentiate yourself from competitors and attract clients with specific needs.
- **Build a portfolio:** Create a portfolio of translated work samples to showcase your skills

and demonstrate your expertise to potential clients.

- **Leverage existing connections:** Reach out to your personal and professional network to let them know you're offering translation services and ask for referrals.

TIPS FOR MAXIMIZING YOUR EARNINGS POTENTIAL

To make the most of your translation side hustle, consider these tips.

- **Develop specialized skills:** Invest in learning specialized skills, such as legal or medical translation, to increase your value to clients and command higher rates.
- **Enhance your language proficiency:** Continuously improve your language skills through formal education, immersion experiences, or self-study to deliver higher-quality translations.
- **Set clear boundaries:** Establish your working hours and availability to maintain a healthy work-life balance while managing your side hustle.

- **Seek long-term clients:** Aim to build long-term relationships with clients, as consistent work can lead to increased earnings and stability in your side hustle.
- **Continuously improve:** Regularly seek feedback from clients and fellow translators to identify areas for improvement and enhance your translation skills.

Language translation services can be a rewarding and profitable side hustle for people with strong language skills and a desire to bridge communication gaps. Anyone can turn their language skills into a profitable side hustle with the help of ChatGPT. You can build a successful side business in the translation industry or the other top 10 side hustles by following the tips and strategies outlined in this book.

119 ESSENTIAL PROMPTS FOR PROFITABLE SIDE HUSTLES

Consider these prompts your shortcut to launching your next side hustle or scaling your current one. Use these prompts in conjunction with the guide *ChatGPT for Cash Flow.*

Be sure to check out our other books to learn even more ways to leverage ChatGPT and artificial intelligence to make more money faster online. You can find these additional resources by going to: https://www.amazon.com/stores/Mark-Silver/author/B0C43T6V7K/

PROMPTS FOR BLOG AND CONTENT WRITING SIDE HUSTLE

- Write a blog post about the benefits of mindfulness meditation for mental health.
- Create a product review for a new line of skincare products and compare them to other popular brands on the market.
- Write a how-to guide for beginners on building a successful social media strategy for small businesses.
- Create a listicle of the top 10 travel destinations to visit in 2023 and why they should be on your bucket list.
- Write a personal essay on the challenges and rewards of working from home in today's digital age.
- Create a buying guide for eco-friendly products and the benefits of choosing sustainable options.
- Write a feature article on the rise of veganism and its impact on the food industry.
- Write an About Us page that tells the story of your company and why you started it.
- Create a landing page for a new product or service that showcases its features and benefits.

- Write an FAQ page that answers common questions about your company, products, and services.
- Create a blog page that highlights your company's expertise and industry knowledge.
- Write a product page that provides detailed information on each of your products, including pricing and specifications.
- Create a testimonial page that showcases customer reviews and feedback.
- Write a page that outlines your company's values and mission statement.
- Create a careers page that lists job openings and company culture.
- Write a contact page that includes a contact form, phone number, and email address for customer inquiries.
- Create a resources page that offers helpful articles, guides, and resources for your audience.

PROMPTS FOR VIDEO SCRIPT AND DESCRIPTION WRITING

- Create a video script and description for a product launch video that showcases the features and benefits of a new fitness tracker.

- Write a video script and description for a brand story video that highlights the history and mission of a sustainable fashion brand.
- Create a video script and description for a tutorial video that teaches viewers how to make a DIY home decor project.
- Write a video script and description for an explainer video that breaks down the science behind climate change and its effects on the environment.
- Create a video script and description for a promotional video that promotes a new travel destination and highlights its unique attractions.
- Write a video script and description for a how-to video that teaches viewers how to cook a new recipe from scratch.
- Create a video script and description for an educational video that explores the history and cultural significance of a specific type of music.
- Write a video script and description for a testimonial video that features real customers sharing their experiences and feedback about a product or service.
- Create a video script and description for an interview video that features an expert in a

specific field sharing their knowledge and insights.
- Write a video script and description for a vlog-style video that documents a day in the life of a professional in a specific industry.

PROMPTS FOR COPYWRITING SERVICE USING CHATGPT

Headlines and Taglines

- Generate headlines and taglines for a new product launch.
- Come up with a catchy tagline for a business slogan.
- Create a compelling headline for a blog post.

Social Media Posts

- Write a social media post promoting a sale or promotion.
- Generate social media content for a product launch.
- Craft a social media post to engage with your client's followers.

Email Campaigns

- Write an email subject line to increase open rates.
- Create an email campaign to promote a new product.
- Generate content for a welcome email for new subscribers.

Landing Pages

- Write copy for a landing page to promote a service.
- Create content for a landing page to drive conversions.
- Generate copy for a landing page to promote a limited-time offer.

PROMPTS FOR SOCIAL MEDIA MANAGEMENT SIDE HUSTLE

1. Generate 10 engaging Instagram caption ideas for a fitness brand.
2. What are the top 5 social media trends for businesses in 2023?
3. Create a 30-day content calendar for a small bakery's Facebook page.

4. What are the best practices for increasing engagement on LinkedIn?
5. How can a travel agency effectively use Pinterest to attract more clients?
6. Draft a series of 5 tweets for a non-profit organization promoting their upcoming charity event.
7. What are the key metrics to track for a successful Instagram marketing campaign?
8. Suggest 3 ways to improve the visual aesthetics of a fashion brand's Instagram feed.
9. How can a real estate agent use social media to generate leads and increase sales?
10. Develop a social media strategy to grow a local restaurant's online presence and attract more customers.
11. What are the most effective techniques for growing a YouTube channel for a tech review business?
12. Generate a list of 10 Facebook post ideas for a veterinary clinic.
13. How can a personal finance blog leverage Twitter to increase audience engagement?
14. What are the key elements of a successful influencer marketing campaign on Instagram?

15. Develop a TikTok content strategy for an eco-friendly cosmetics brand to connect with a younger audience.
16. Suggest 5 ways to optimize a LinkedIn company page for better visibility and engagement.
17. How can a podcast use social media platforms to increase its listenership?
18. Create a list of 10 Instagram Story ideas for a motivational speaker to inspire their followers.
19. What are the best practices for handling negative comments and feedback on social media?
20. How can a local bookstore use social media to create a sense of community and drive in-store traffic?

PROMPTS FOR WRITING RESUMES AND BIOS

1. Generate a compelling professional summary for a marketing manager with 5 years of experience.
2. Create a list of 10 action verbs to effectively describe accomplishments in a resume.
3. Draft a LinkedIn bio for a recent college graduate with a degree in computer science.

4. Suggest 5 resume-formatting tips to create a visually appealing and easy-to-read document.
5. Write an engaging personal bio for a freelance graphic designer's website.
6. How can transferable skills be effectively showcased in a career change resume?
7. Create a list of 5 questions to ask a client before writing their resume or bio.
8. Develop a powerful resume objective statement for a sales professional targeting a managerial role.
9. Provide tips for optimizing a LinkedIn profile to increase visibility and attract recruiters.
10. Write an attention-grabbing personal bio for an aspiring author's social media platforms.

PROMPTS FOR VIRTUAL ASSISTANCE AS A SIDE HUSTLE

1. List 10 essential skills that every successful virtual assistant should possess.
2. What are the most effective strategies for managing multiple clients and tasks as a virtual assistant?
3. Describe the process of setting up a professional virtual assistant website to attract potential clients.

4. How can a virtual assistant use social media to expand their network and market their services effectively?
5. Discuss the benefits and challenges of working as a virtual assistant from a remote location.
6. Provide a list of 5 tools and apps that can help virtual assistants stay organized and productive.
7. What are the key factors to consider when pricing virtual assistant services?
8. How can a virtual assistant establish a strong working relationship with clients and ensure clear communication?
9. Describe the process of identifying a niche or specialty as a virtual assistant and its potential impact on earnings.
10. Outline the steps to create an effective marketing plan for a virtual assistant side hustle.

PROMPTS FOR WRITING AND SELLING CHILDREN'S BOOKS

1. What are the key differences between traditional publishing and self-publishing children's books, and which route is best for a side hustle?

2. How can an aspiring children's book author determine the appropriate age group and reading level for their story?
3. List 10 popular themes and genres in children's literature to inspire story ideas.
4. Discuss the process of finding and collaborating with an illustrator for your children's book.
5. What are the essential elements of a successful children's book marketing campaign?
6. Explain the role of literary agents in the traditional publishing process and how to find the right agent for your children's book.
7. Provide a step-by-step guide for self-publishing a children's book on Amazon Kindle Direct Publishing.
8. How can authors use social media and other online platforms to build a loyal readership and promote their children's books?
9. Discuss strategies for creating diverse and inclusive characters in children's literature.
10. What are the benefits and challenges of writing a series of children's books? How can these benefits and challenges impact earnings potential?

PROMPTS FOR ONLINE SURVEY PARTICIPATION

1. How can you identify high-quality survey platforms and avoid scams when looking for online survey opportunities?
2. What strategies can help you stay motivated and consistent in your online survey participation side hustle?
3. Discuss the importance of honesty and consistency in your survey responses and how it affects your reputation as a survey participant.
4. What are the best practices for managing your time and staying organized when participating in multiple online surveys?
5. How can you effectively handle open-ended survey questions to provide valuable and actionable feedback to researchers?
6. Describe the process of using ChatGPT to generate ideas and opinions when you're unsure how to respond to a survey question.
7. What are some common mistakes to avoid when participating in online surveys to ensure a positive experience and maximize earnings?
8. Discuss the role of demographics and interest information in receiving tailored survey

invitations and how to optimize your profile for better opportunities.

9. How can you balance online survey participation with other side hustles to diversify your income and achieve greater financial stability?
10. Provide tips on creating a comfortable and efficient workspace for online survey participation to maximize productivity and minimize distractions.

PROMPTS FOR ONLINE CUSTOMER SUPPORT AS A SIDE HUSTLE

1. What are the best practices for providing excellent online customer support in a chat-based environment?
2. How can you effectively manage your time and multitask when handling multiple customer inquiries simultaneously?
3. Discuss strategies for developing a personal and professional rapport with customers during online interactions.
4. What are the main differences between providing customer support via chat, email, and phone, and how can you adapt your communication style accordingly?

5. How can you effectively handle language barriers and cultural differences when providing online customer support to an international clientele?
6. Describe the process of tracking and measuring key performance indicators (KPIs) in online customer support to ensure consistent service quality.
7. What are some common customer support scenarios, and how can you prepare for them to provide quick and efficient assistance?
8. Discuss the importance of staying up-to-date with company policies, product updates, and industry news to provide accurate and relevant support to customers.
9. How can you manage and reduce stress in a high-pressure online customer support environment?
10. Provide tips for developing a growth mindset and continuously learning and improving as an online customer support representative.

PROMPTS FOR LANGUAGE TRANSLATION SERVICES

1. What are the key skills and qualities required for success in the language translation services industry?
2. Discuss the importance of cultural awareness and sensitivity when providing language translation services.
3. Explain the differences between translation, localization, and transcreation and how to determine which service is most appropriate for a given project.
4. Describe the process of creating a strong portfolio to showcase your translation skills and attract potential clients.
5. How can you stay up-to-date with language trends, industry news, and best practices in the translation field?
6. What are some common challenges faced by freelance translators, and how can you overcome them?
7. Discuss the role of translation software and tools in the translation process and how to strike a balance between human expertise and technological assistance.

8. How can you effectively manage your time and workload when juggling multiple translation projects and deadlines?
9. What strategies can help you build a strong network of clients and fellow translators to support your side hustle?
10. Provide tips for negotiating rates and contracts with clients to ensure fair compensation and clear expectations for your translation services.

Let Others Know About ChatGPT's Specific Abilities and Tasks So They Can Build a Brighter Financial Future

You have now discovered 10 side hustles in which ChatGPT can help you excel. You know the right prompts to use, how to find clients, and how to maximize your earnings.

Head over to Amazon and let other readers know how this book has helped you. Just one or two sentences could be enough to help them realize their full money-making potential.

Thank you for helping me on my quest to help others understand how to leverage the money-making potential of ChatGPT. I hope you can inspire them to formulate the very first question they plan to ask this invaluable language assistant.

FINAL WORDS

ChatGPT For Cash Flow: *10 Easy Ways to Unlock the Power of AI to Build a Side Hustle Empire & Make Money Online Fast* is an in-depth guide that offers valuable insights into the world of side hustles and how leveraging AI-powered language models can be a game changer.

Dedication, continuous improvement, and a commitment to providing accurate and contextually appropriate translations are the keys to success. You, too, can become a Side Hustle Hero with these qualities and ChatGPT by your side!

To succeed with ChatGPT side hustles, there are a few key things to keep in mind:

1. **Continuously improve your skills:** As with any business, the quality of your work will determine your success. Continuously honing your language and writing skills will help you stand out from the competition.
2. **Provide accurate and contextually appropriate translations:** This is crucial for building a strong reputation and gaining repeat clients. Always ensure that your translations are not only accurate but also take into account the cultural nuances and context of the content.
3. **Utilize ChatGPT's capabilities to the fullest:** ChatGPT is a powerful tool that can help you automate certain tasks and save time. Be sure to take advantage of its capabilities to streamline your workflow and increase efficiency.
4. **Focus on marketing and networking:** Finding clients is vital to success in any business, and the same is true for ChatGPT side hustles. Utilize social media, online marketplaces, and networking events to connect with potential clients and build your brand.

By following these tips and committing to providing high-quality work, you can build a successful side business with ChatGPT. Remember that success may not come overnight, but with dedication and hard work, you, too, can become a Side Hustle Hero!

ABOUT THE AUTHOR

Mark Silver (aka *The AI Entrepreneur*) is a serial entrepreneur, investor, and innovator. His mission is to make people wealthy by sharing knowledge, information, investment opportunities, and new ways to make money online with one million people.

By changing their lives, Mark can impact billions of people. An obstacle for many people is money. He plans to remove that burden with his books, courses, and programs so more people can focus on their true purpose and passions.

ALSO BY MARK SILVER

The AI Entrepreneur Series

ChatGPT for Cash Flow

ChatGPT for Side Hustles

ChatGPT for Home-Based Businesses

ChatGPT for Authors

ChatGPT for Book Outlines

ChatGPT for Copywriters

ChatGPT for Bloggers

ChatGPT for Content Creators

Made in United States
Troutdale, OR
04/11/2024